Earth
Our Big Home!

An imprint of Om Books International

Bob was in his classroom, sitting next to his friend Nina. He was looking forward to his class. Bob loved studying. "Good Morning!" Miss Kelly greeted the children with a smile. "Today, we are going to learn about the Earth."

"Yes!" exclaimed Bob, happily.

"What is so interesting about the Earth? Why are you so excited?" Nina asked him.

Miss Kelly who had overheard their conversation smiled and turned to the class, "Can anyone tell me what is so special about our planet Earth?" she asked them.

Bob's hand shot up in the air. He seemed to have an answer to everything.

"Earth is a planet. It is the only planet where humans live!"

"What is a planet, Miss?" Nina asked.

Miss Kelly showed them a colourful model of the solar system. She asked all the children to gather around her table.

"Planets are large, natural objects that travel around stars," explained Miss Kelly, pointing towards the model. "The Sun is a star and eight planets move around it."

"Miss, why is Earth the only planet in the solar system that supports life?" asked Mary.

"That's because the other planets are either too hot or too cold. Earth, on the other hand, has an atmosphere that supports the right temperature or else we would all have been burnt or frozen," Bob said.

"That's right, Bob! This blue ball here is Earth. Just like the ball, Earth is round. As 71% of Earth is covered with water, it looks blue from space. Earth is the only planet that has oxygen and that is what makes life possible here," Miss Kelly said.

Miss Kelly led the children to the playground. The children gathered around her, curious to know why she had brought them there.

"Bob, come here and hold this yellow ball," said Miss Kelly, asking Bob to stand in the middle of the playground.

"Bob is the Sun. The Sun is at the centre of the solar system."

She gave the blue ball to Nina.

"Nina, you are the Earth. You have to move around Bob. Children, this is how all the planets revolve around the Sun. We have 365 days in one year because that is the time it takes the Earth to complete one revolution around the Sun."

"Oh! Is that why we have day and night on Earth, Miss?" Mary asked.

Miss Kelly laughed. "No Mary. It is due to the rotation of the Earth that we have day and night. The side of the Earth which faces the Sun has day, while the side which faces away from the Sun has night," Miss Kelly explained.

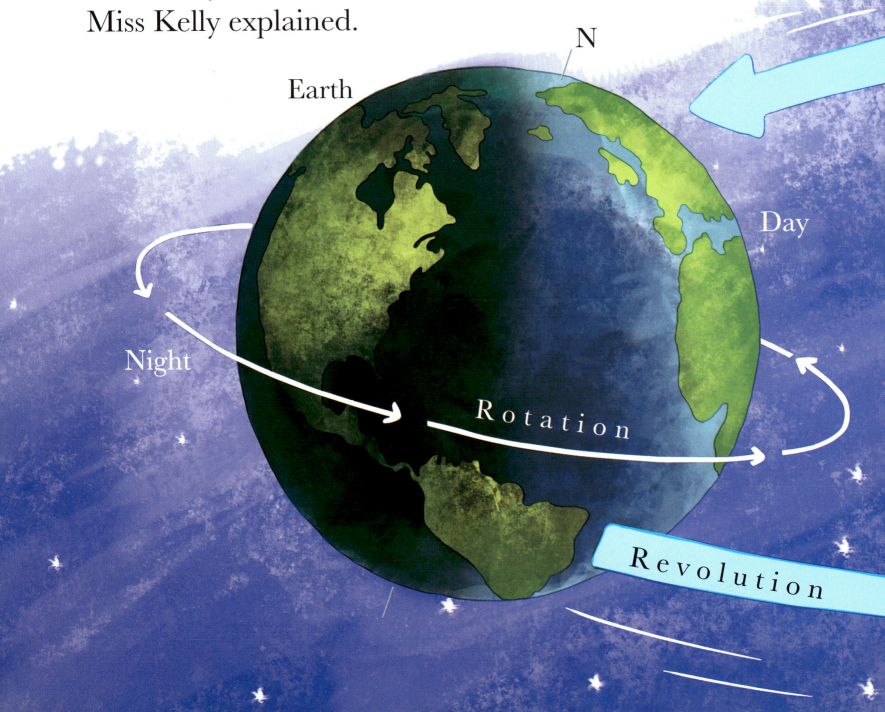

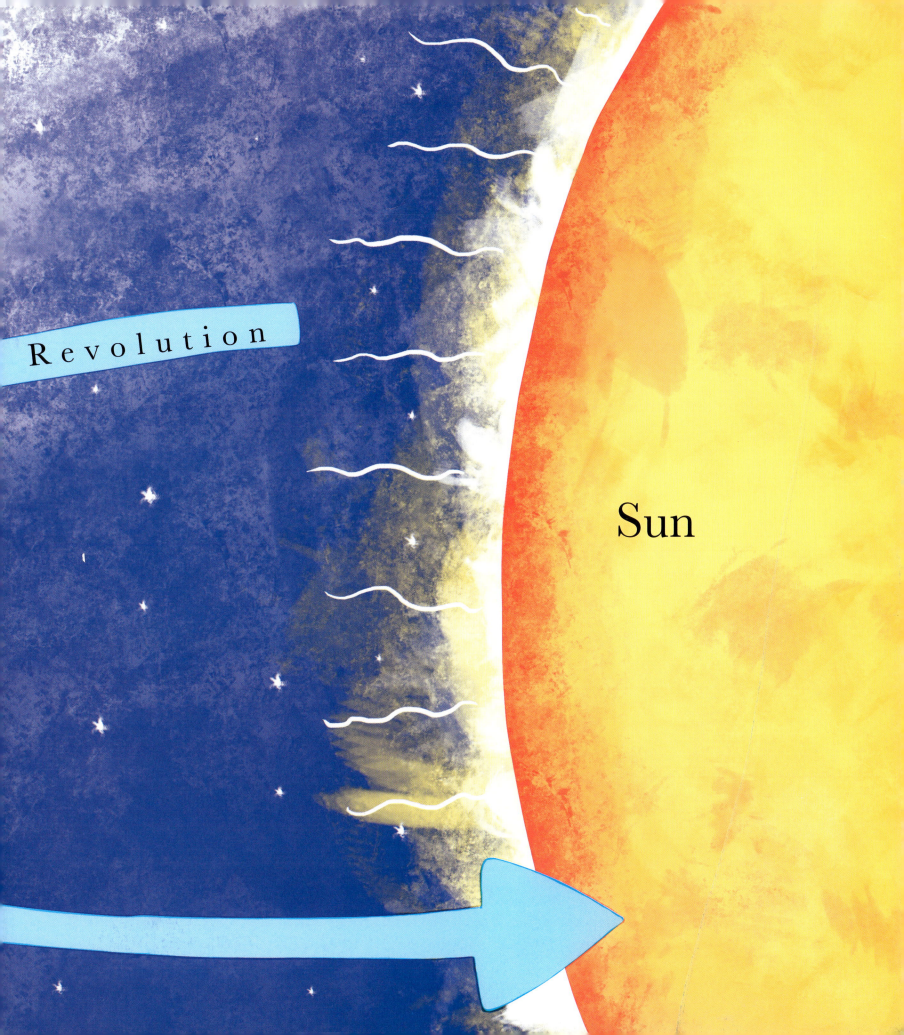

"Why are there seasons on Earth?" Sam asked.

"We have different seasons because Earth is tilted on its axis. This means that the Earth doesn't rotate in a straight line. When the northern part of the Earth is tilted towards the Sun, it experiences summer. Meanwhile, the southern half which is tilted away from the Sun has winter, and vice versa.

Can anyone name all the seasons?" Miss Kelly asked.

Nina raised her hand and said, "I can! There are four seasons. They are spring, summer, autumn and winter."
"Very good, Nina!"

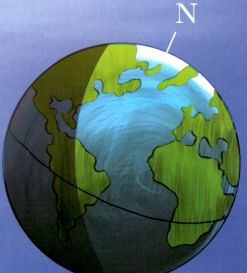

Spring

Summer

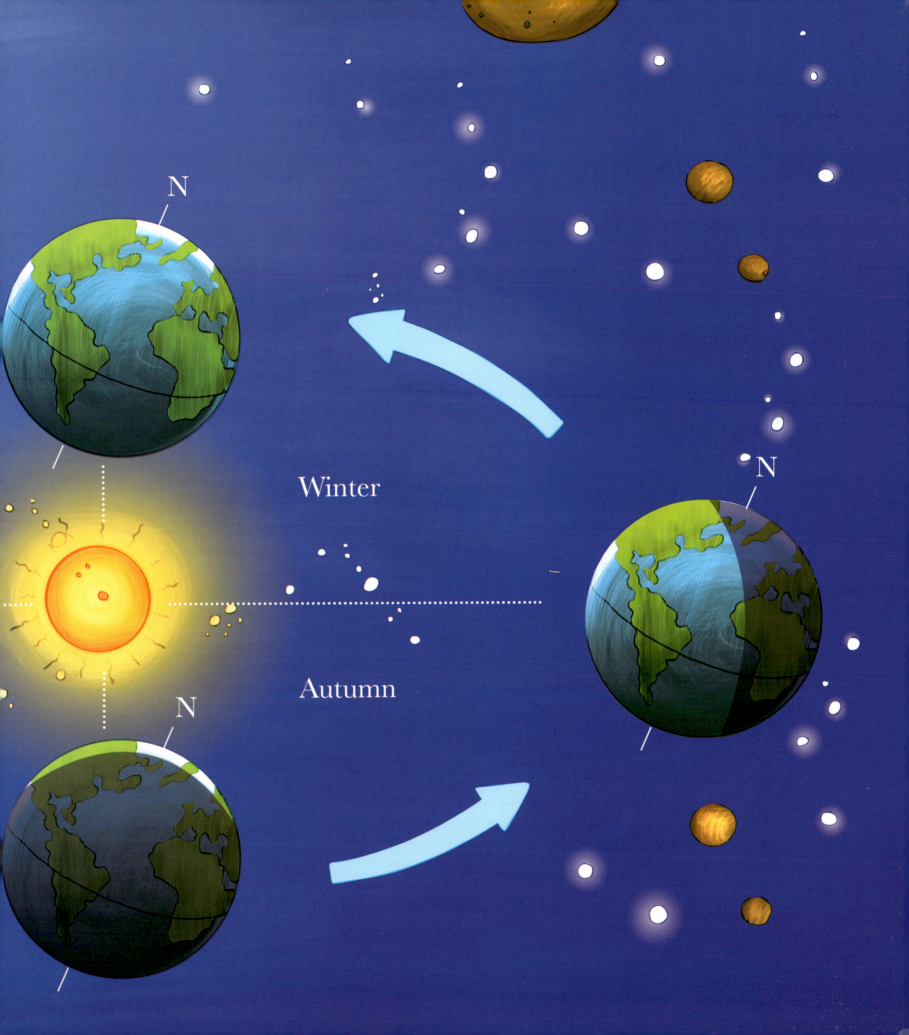

"But do you know that Earth is in danger?" She asked.

"Why?" Bob asked, shocked.

"The Earth helps us to live. It's friendly. But we don't respect it. We are making the Earth's air, water and land very dirty. We call it pollution. We also harm its best friend, the Ozone layer that protects us from harmful ultraviolet rays of the sun. The Earth is getting sicker day by day. For that reason, the beautiful seasons are losing their flavour too."

Spring

Summer

Autumn

Winter

"It's because we are dirtying the air, water and land!" Mary exclaimed.

"Exactly! Human activities are destroying the Earth. The smoke from factories and vehicles pollutes the air. The waste we dump on land and in water harms the environment. The forests are disappearing because we are cutting down trees for our selfish needs. All this is damaging the Earth," Miss Kelly said.

"That is so wrong!" Nina said, sadly.

"If we continue harming our planet, soon humans won't be able to live here. Would you want that to happen?"

"No, Miss. We will protect the environment. We love our planet!" the children said in unison.